Garden landscape Detail Design - 4

庭院细部元素设计

汀步、小桥、车档、座椅、标志牌、雕塑小品

④

中国林业出版社
China Forestry Publishing House

图书在版编目（CIP）数据

庭院细部元素设计 . ④，汀步、小桥、车档、座椅、标志牌、雕塑小品 /《庭院细部元素设计》编委会编 . —— 北京：中国林业出版社，2016.5

ISBN 978-7-5038-8503-7

Ⅰ . ①庭… Ⅱ . ①庭… Ⅲ . ①庭院 - 景观设计 Ⅳ . ① TU986.4

中国版本图书馆 CIP 数据核字 (2016) 第 078633 号

《庭院细部元素设计》编委会

◎ 编委会成员名单

主　编：董　君

编写成员：董　君　李晓娟　曾　勇　梁怡婷　贾　濛　李通宇　姚美慧
　　　　　刘　丹　张　欣　钱　瑾　翟继祥　王与娟　李艳君　温国兴
　　　　　黄京娜　罗国华　夏　茜　张　敏　滕德会　周英桂　李伟进

◎ 特别鸣谢：北京吉典博图文化传播有限公司

中国林业出版社 · 建筑与家居出版中心

责任编辑：纪　亮　王思源
联系电话：010-8314 3518

出版：中国林业出版社
　　　（100009 北京西城区德内大街刘海胡同 7 号）
http://lycb.forestry.gov.cn/
E-mail：cfphz@public.bta.net.cn
电话：（010）8314 3518
发行：中国林业出版社
印刷：北京利丰雅高长城印刷有限公司
版次：2016 年 6 月第 1 版
印次：2016 年 6 月第 1 次
开本：235mm×235mm 1/12
印张：16
字数：200 千字
定价：99.00 元

鸣谢
因稿件繁多内容多样，书中部分作品无法及时联系到作者，请作者通过编辑部与主编联系获取样书，并在此表示感谢。

目录 4
CONTENTS

汀步 5~52

小桥 53~94

车档 95~98

座椅 99~222

标志牌 223~234

雕塑小品 235~320

GARDEN LANDSCAPE DETAIL DESIGN · STEPPING

汀步

庭院的意义在于居住者对它所提供的生活方式的认同，在于它所带来的院落"能指"（居住环境）和"所指"（居住者的生活方式、社会交往方式）的和谐关系，而此代码在现代集合住宅中的被扭曲和遗忘引发了居民对院落中亲密的邻里关系的怀念。

庭院作为人为的自然空间，它将自然和人工相结合起来，让人产生一种回归自然的向往。庭院中往往常采用小巧、精致的手法，设置植物、院路、亭廊、假山、雕塑、水池等，既表现出个性化的特点，同时给人们带来自然美、人工美的艺术享受。

Agnatus eiumendae voluptamusa custet et demporeribus minverchil mi, unte magnit eum et que doloreh enihili quaspe vendi ium nume sum vita doloria nosam as utendi te solori beribus pa consequi idias maiorem porestiis exerendias as es ut volupit, el earum labo. Itatemp orerenes ipsunt, simusap idustenet explacea et. con rat doluptatus arum et que plit occum et iunt quam, voluptia parcius dollect aturehentum abo. Neque voluptas seque sam, qui opta volorib ernate lat latem eumet la quis qui ut arit, am reperates dolore.

1-3.Garden in the Silver Moonlight
4，5.Greenwich Residence, Greenwich, Connecticut

Stepping ○ 汀步

1. Austin, USA Date of Completion-串联式溪边住宅
2. Darwin Waterfront Public Domain
3.28
4，5.Barcelona, Spain Date of Completion-Tanatorio Ronda de Dalt花园
6-8.Bassil Mountain Escape, Faqra, Lebanon

Stepping ○ 汀步

Stepping ○ 汀步

1. Bassil Mountain Escape, Faqra, Lebanon
2. Charleston, USA Date of Completion-查尔斯顿滨水公园
3. copyright ADEPT SCHONHERR downtown-宽容之城
4，5. City Park, Bradford
6. FZK_021
7. FZK_023
8. Garden in the Silver Moonlight

Stepping 汀步

1. Holstebro, Denmark—If it is outstanding, celebrate it twice
2. Houston, TX—ConocoPhillips World Headquarters
3. Private Residence, San Francisco, California
4. Malibu Beach House, Malibu, California
5. Lunada Bay Residence, Palos Verdes Peninsula, Southern Californi

1.PA-323-012
2.Shreveport-雨花园别墅
3.Pamet Valley
4.Sonoma Vineyard, Glen Ellen, California
5.Villa H. St. Gilgen
6.Washington-2200宾夕法尼亚大道

Stepping ○ 汀步

1.阿姆斯特丹-辛克尔小岛
2.Xochimilco, Mexico City.-XOCHIMILCO ECOLOGICAL PARK
3.Speckman House Landscape, Highland Park, St. Paul, MN
4-6.埃布罗河畔
7.爱尔兰都柏林-观海

Stepping ○ 汀步

1-3.安徽省-合肥政务文化主题公园
4，5.澳大利亚-库吉港
6.澳大利亚悉尼-Bondi 至 Bronte 滨海走廊

Stepping ○ 汀步

Stepping ○ 汀步

1-3.芭堤雅希尔顿酒店
4.半岛湖边小屋
5,6.澳大利亚悉尼-悉尼5号湿地

Stepping ○ 汀步

1，2.达尔文海滨公共区域
3.德国巴伐利亚02007 Waldkirchen花园展
4，5.袋鼠湾
6.佛山-天安鸿基花园-天安b
7.广州-保利国际广场
8.翡翠岛

1.观塘别墅庭院
2.杭州－江南水乡碧水铭苑
3.广州－凤凰城8号
4.广州－华南碧桂园
5，6.广州－凤凰城8号

Stepping ○ 汀步

Stepping ○ 汀步

1. 荷兰阿姆斯特丹-Floriande居住区
2. 韩国首尔-West Seoul Lake Park
3，4. 华南碧桂园燕园
5. 河源城区-河源东江·首府花园

1-3.加拿大，私人公寓，兰乔圣菲
4-7.加拿大-东南福溪可持续社区

Stepping ○ 汀步

Stepping ○ 汀步

1. 杰克·埃文斯船港—堤维德岬一期工程
2. 江南华府
3. 加拿大-东南福溪可持续社区
4. 金碧湖畔6号
5. 乐山市-恒邦·翡翠国际社区
6. 绿洲千岛花园
7. 美国加利福尼亚-特伦顿大道
8. 荔湖城
9. 美国纽约-亚瑟·罗斯平台

Stepping ○ 汀步

.墨尔本 – Frankston
.墨西哥 – 高科技办公园区 – Tecnoparque
rancisco Gez Sosa – 41
.墨西哥 – 马里纳尔可住宅136-03
–6.南海市民广场和千灯湖公园

1,2.南海市民广场和千灯湖公园
3.纽约-龙门广场州立公园
4.纽约-威廉斯堡滨水演出会场02
5.挪威-Gudbrandsjuvet -Viewing platforms && bridges

Stepping ○ 汀步

1-4.挪威奥斯陆-Rolfsbukta住宅区
5.七宝
6.挪威-南森公园
7.日式枯山水庭院
8.桑德贝滨水景观，亚瑟王子码头

Stepping ○ 汀步

Stepping ○ 汀步

1.萨缪尔·德·尚普兰滨水长廊
2，3.上海-底楼花园
4.萨缪尔·德·尚普兰滨水长廊
5.上海-绿洲江南园
6.上海市-观庭

1.上海市-花语墅
2-4.上海市-金地格林
5.上海市-新律花园
6.上海市-平江路底楼花园
7.上海市-太阳湖大花园设计造园

Stepping ○ 汀步

1. 上海－太阳湖大花园设计造园
2. 上海－香溪澜院
3，4. 上海－汤臣高尔夫
5. 佘山－3号园
6. 顺德－碧桂园
7. 布卡克海滩

1.深圳-SHENZHEN BAY COASTLINE PARK
2.佘山－东紫园
3.圣淘沙跨海步行道
4，5.斯塔斯福特
6.台湾－南投农舍别墅－waterfall rest area
7.太湖高尔夫
8.泰国曼谷-Prive by Sansiri
9.唐山－唐人起居－w02

Stepping ○ 汀步

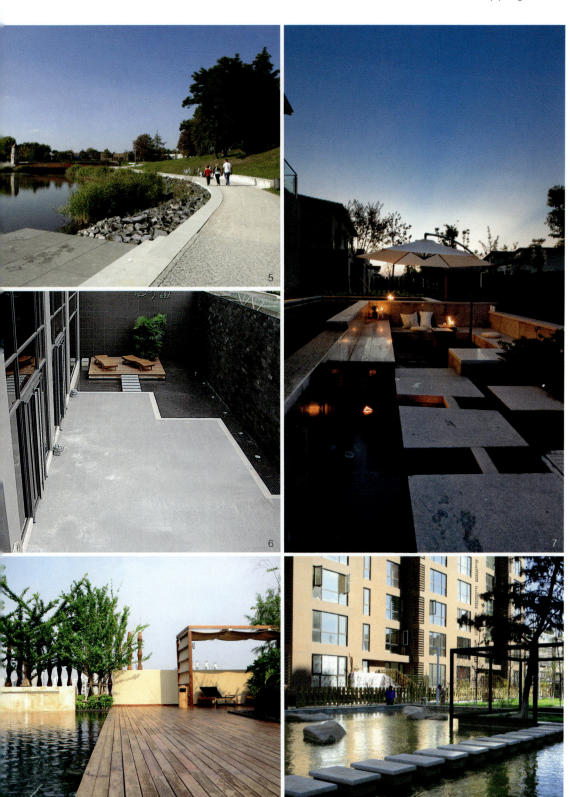

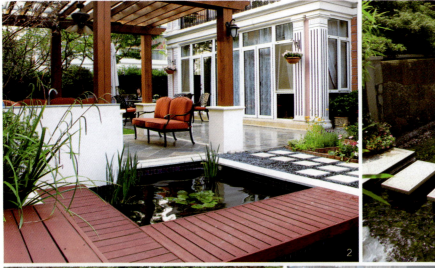

Stepping ○ 汀步

1，2.张家港 怡佳苑
3.上海－绿洲江南
4.重庆－梦中天地
5.观塘别墅庭院
6.文华别墅
7.威尔克斯巴里堤防加高河流公地
8.西郊花园
9.西班牙圣塞瓦斯蒂安-AITZ TOKI 别墅

Stepping ○ 汀步

1. 悉尼-Pirrama公园
2. 西郊一品
3. 西雅图-山顶住宅
4. 雅居乐私家花园
5. 阳光九九造园
6. 万科朗润园

Stepping ○ 汀步

1，2.叶片小斋leaf house
3.月湖山庄假山水景 – 月湖细节
4，5.银都名墅
6.云栖蝶谷

1.挪威奥斯陆-Rolfsbukta住宅区
2.澳大利亚悉尼-悉尼5号湿地
3.雅居乐私家花园

GARDEN LANDSCAPE DETAIL DESIGN · SMALL BRIDGE

小桥

庭院的意义在于居住者对它所提供的生活方式的认同，在于它所带来的院落"能指"（居住环境）和"所指"（居住者的生活方式、社会交往方式）的和谐关系，而此代码在现代集合住宅中的被扭曲和遗忘引发了居民对院落中亲密的邻里关系的怀念。

庭院作为人为的自然空间，它将自然和人工相结合起来，让人产生一种回归自然的向往。庭院中往往常采用小巧、精致的手法，设置植物、院路、亭廊、假山、雕塑、水池等，既表现出个性化的特点，同时给人们带来自然美、人工美的艺术享受。

gnatus eiumendae voluptamusa custet et demporeribus minverchil mi, unte magnit eum et que doloreh enihili uaspe vendi ium nume sum vita doloria nosam as utendi te solori beribus pa consequi idias maiorem porestiis xerendias as es ut volupit, el earum labo. Itatemp orerenes ipsunt, simusap idustenet explacea et. on rat doluptatus arum et que plit occum et iunt quam, voluptia parcius dollect aturehentum abo. Neque voluptas eque sam, qui opta volorib ernate lat latem eumet la quis qui ut arit, am reperates dolore.

1. Canada-cSimcoe WaveDeck
2. Expo 2008 Main Building
3. Holstebro, Denmark-If it is outstanding, celebrate it twice
4. Jeonnam-小岛上的动物园--JDS_DOCHODO ISLAND ZOO - AVIARY INTERIOR 02
5，6. Madrid RIO
7. Linz, Austria-乡村别墅公园和散步广场

Small bridge ○ 小桥

Small bridge ○ 小桥

2.Shreveport－雨花园别墅
Sitges－肯·罗伯特公园2110 39 copia 3
Unfolding Terrace, Dumbo, Brooklyn, New York
6.阿姆斯特丹－辛克尔小岛
埃布罗河畔

57

Small bridge ○ 小桥

1，2.安徽省-合肥绿地国际花都二期"玫瑰苑"
3，4.安斯伯利镇区

1,2.成都－玉都别墅庭园
3.百家湖
4,5.北京－泰禾运河岸上的院子
6,7.波特兰花园60号
8.成都－翡翠城汇锦云天4

Small bridge ○ 小桥

1.成都-紫檀山
2.袋鼠湾
3-5.德国巴伐利亚02007 Waldkirchen花园展
6,7.德国柏林-晋城市儿童公园

Small bridge ○ 小桥

Small bridge ○ 小桥

1.俄勒冈州俄勒冈市-乔恩斯托姆公园
2.帝景天成坡地别墅
3.东莞市-湖景壹号庄园私家庭院
4.翡翠岛
5.法国克洛尔-Renewal of the main street
6，7.佛山-天安鸿基花园

1. 广州-保利国际广场
2. 广东佛山-山水庄园高档别墅花园
3. 广州-凤凰城8号
4. 广州-华南碧桂园
5. 广州-汇景新城别墅

Small bridge ○ 小桥

1. 荷兰海牙-playground Melis Stokepark
2-4.霍恩博格海滨公园
5，6.Adelaide Zoo Giant Panda Forest
7.Darwin Waterfront Public Domain

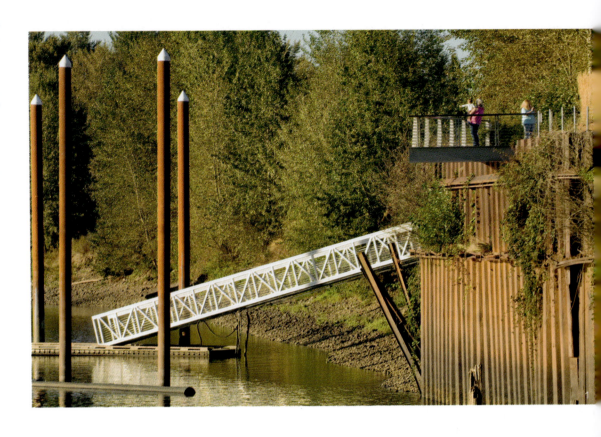

Small bridge ○ 小桥

1.美国俄勒冈-Jon Storm Park
2,3.美国普罗维登斯-The Steel Yard
4,5.墨尔本 - Frankston - DSCF0659-2
6.荷兰Heerhugowaard-South-Park of Luna
7.墨尔本-儿童艺术游乐园

1,2.墨尔本-皇家湿地公园
3,4.南海市民广场和千灯湖公园
5.纽约-龙门广场州立公园
6,7.挪威-Gudbrandsjuvet -Viewing platforms && bridges

Small bridge ○ 小桥

Small bridge ○ 小桥

1.湖景壹号别墅庄
2.绿洲千岛花园
3.江苏－睢宁流云水袖桥
4.霍恩博格海滨公园
5.绿洲千岛花园

Small bridge ○ 小桥

1. 挪威-南森公园
2，3.七宝
4.泉州-奥林匹克花园别墅区67-泳池景观
5.桑德贝滨水景观，亚瑟王子码头
6.上海-底楼花园
7.华南碧桂园燕园
8.荷兰阿姆斯特丹-"Meer-park" Amsterdam

Small bridge ○ 小桥

1.上海－绿洲江南园
2.上海市－大华锦绣
3.上海市－东町庭院
4.上海市－比利华1
5.上海市－花语墅

1.上海市-金地格林
2，3.上海市-平江路底楼花园
4，5.上海市-汤臣高尔夫别墅
6.荷兰阿姆斯特丹-"Meer-park" Amsterdam
7.上海市-月湖山庄假山水景
8.上海-太阳湖大花园设计造园

Small bridge ○ 小桥

Small bridge ○ 小桥

1-4.上海－香溪澜院
5.上海－银都名墅-花园

Small bridge ○ 小桥

1，2.上海－银都名墅－花园
3-6.深圳-SHENZHEN BAY COASTLINE PARK
7.沈阳市－汇程庭院
8.沈阳－万科金域蓝湾4

Small bridge ○ 小桥

1，2.沈阳-万科金域蓝湾
3.圣地亚哥-拉卡斯塔格伦社区
4.顺德－碧桂园

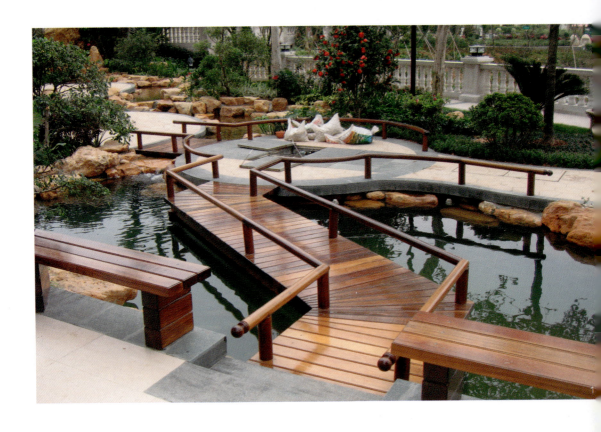

1，2.顺德－碧桂园
3.台湾-基隆海洋广场
4.台湾商会
5.天鹅湖
6-8.特拉福德滨河长廊

Small bridge ○ 小桥

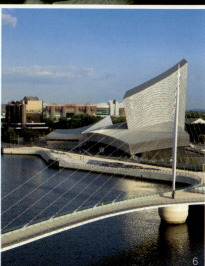

Small bridge ○ 小桥

2.天津－桥园公园
4.武汉－天下别墅
威尼斯水城别墅花园nEO

1. 新港码头
2. 西郊大公馆107号庭院
3. 新太阳国际养生城
4. 新西兰-杰利科北部码头漫步长廊，杰利科大道和筒仓公园
5，6. 亚洲大酒店
7. 易郡20090708015

Small bridge ○ 小桥

1.棕榈湾
2.重庆－梦中天地
3，4.钻石提格公园

GARDEN LANDSCAPE DETAIL DESIGN · THE CAR STALLS

车档

　　庭院的意义在于居住者对它所提供的生活方式的认同，在于它所带来的院落"能指"（居住环境）和"所指"（居住者的生活方式、社会交往方式）的和谐关系，而此代码在现代集合住宅中的被扭曲和遗忘引＿了居民对院落中亲密的邻里关系的怀念。

　　庭院作为人为的自然空间，它将自然和人工相结合起来，让人产生一种回归自然的向往。庭院中往往常＿用小巧、精致的手法，设置植物、院路、亭廊、假山、雕塑、水池等，既表现出个性化的特点，同时给人们带来自然美、人工美的艺术享受。

gnatus eiumendae voluptamusa custet et demporeribus minverchil mi, unte magnit eum et que doloreh enihili uaspe vendi ium nume sum vita doloria nosam as utendi te solori beribus pa consequi idias maiorem porestiis xerendias as es ut volupit, el earum labo. Itatemp orerenes ipsunt, simusap idustenet explacea et.
on rat doluptatus arum et que plit occum et iunt quam, voluptia parcius dollect aturehentum abo. Neque voluptas eque sam, qui opta volorib ernate lat latem eumet la quis qui ut arit, am reperates dolore.

1.墨西哥－高科技办公园区－Tecnoparque Francisco Gez Sosa
2.加利福尼亚－马里布海岸豪宅
3.Hilltop Residence, Seattle, WA
4，5.上海市－汤臣高尔夫别墅
6.云间绿大地

The car stalls ○ 车档

1.Hilltop Residence, Seattle, WA
2，3.尊

ARDEN LANDSCAPE DETAIL DESIGN · CHAIR

坐椅

　　庭院的意义在于居住者对它所提供的生活方式的认同，在于它所带来的院落"能指"（居住环境）和"所指"（居住者的生活方式、社会交往方式）的和谐关系，而此代码在现代集合住宅中的被扭曲和遗忘引起了居民对院落中亲密的邻里关系的怀念。

　　庭院作为人为的自然空间，它将自然和人工相结合起来，让人产生一种回归自然的向往。庭院中往往常用小巧、精致的手法，设置植物、院路、亭廊、假山、雕塑、水池等，既表现出个性化的特点，同时给人们带来自然美、人工美的艺术享受。

gnatus eiumendae voluptamusa custet et demporeribus minverchil mi, unte magnit eum et que doloreh enihili uaspe vendi ium nume sum vita doloria nosam as utendi te solori beribus pa consequi idias maiorem porestiis xerendias as es ut volupit, el earum labo. Itatemp orerenes ipsunt, simusap idustenet explacea et.
on rat doluptatus arum et que plit occum et iunt quam, voluptia parcius dollect aturehentum abo. Neque voluptas eque sam, qui opta volorib ernate lat latem eumet la quis qui ut arit, am reperates dolore.

Chair ○ 座椅

1._上海－圣安德鲁斯庄园
2-5.8a The Terrace
6.Anchorage, Alaska, USA-Anchorage Museum Expansion
7.Alameda, USA Date of Completion-海湾船艇公司
8.Adelaide Zoo Giant Panda Forest

1, 2.Adelaide Zoo Giant Panda Forest
3.Aley, Lebanon Date of Completion-消失点
4, 5.Austin, USA Date of Completion-串联式溪边住宅
6, 7.Australia-Adelaide Zoo Entrance Precinct

Chair ○ 座椅

Chair ○ 座椅

1.Bassil Mountain Escape, Faqra, Lebanon
2.Canada-cSimcoe WaveDeck
3.Castell D'emporda露天酒店
4-6.Australia-Adelaide Zoo Entrance Precinct

Chair ○ 座椅

-4.Berlin-VISITOR
,6.Berlin－－三角铁路站公园－－Com_Loidl_BerlinPARK
.Connecticut Country House, Westport, Connecticut

Chair ○ 座椅

.DPP_37499
2-5.Corner of Hay and William Streets, Perth, WA-
Wesley Quarter
6.DSC00481

109

1，2.Faqra－魔力居所
3-6.Genk Belgium-m!ne
7.Eugene, Oregon, USA-John E. Jaqua Academic Center for Students Athletes

Chair ○ 座椅

111

1.Houston, TX-ConocoPhillips World Headquarters
2.Lima－绿意来袭
3.Houston, TX-ConocoPhillips World Headquarters
4.l上海－新律花园
5.Horizon Residence, Venice, California

1. Hilltop Residence, Seattle, WA
2. Holstebro, Denmark-If it is outstanding, celebrate it twice
3. Horizon Residence, Venice, California
4-7. Linz, Austria-乡村别墅公园和散步广场
8. Los Angeles, CA-Wilmington Waterfront Park

Chair ○ 座椅

Chair ○ 座椅

1.Madrid RIO
2.Melboume－圣基尔达海滨城的交通线
3.Manhattan Roof Terrace, New York, New York
4.PA-323-011
5，6.New York-Beach House, Amagansett

1.PA-323-015
2.Pamet Valley
3.Plenty Road, Bundoora, Victoria-University Hill Town Centre && Wetland
4.Private Residence, San Francisco, California
5.Private ResidenceGarden of Planes, Richmond, VA

Chair ○ 座椅

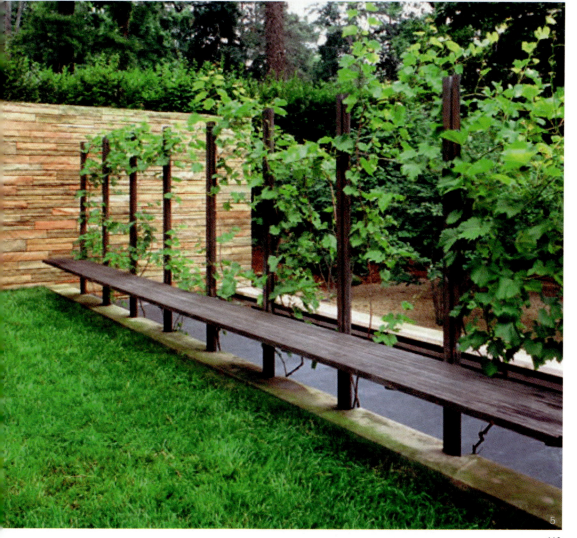

Chair ○ 座椅

1.Quartz Mountain Residence, Paradise Valley, Arizona
2，3.Private ResidenceGarden of Planes, Richmond, VA
4，5.Rotterdam－医院办公楼景观－Maasstad Hospital 09－seating elements by night

1.San Luis Potosí, Mexico Date of Completion-圣路易
总体规划
2.Sitges－肯・罗伯特公园12 seure
3，4.Speckman House Landscape, Highland Park, St. Paul, MN
5.Stanislaw Lem memorial – Garden of Experience - Educational Park in Krakow

Chair ○ 座椅

1.Stanislaw Lem memorial – Garden of Experience – Educational Park in Krakow
2.Tables of Water, Lake Washington, Washington
3.Stanislaw Lem memorial – Garden of Experience – Educational Park in Krakow
4，5.The Hague, The Netherlands–park Vlaskamp
6.Tolosa (Gipúzcoa)–Jolastoki (tolosa)
7.爱尔兰都柏林–登陆

Chair ○ 座椅

1.爱尔兰都柏林-登陆
2.Unfolding Terrace, Dumbo, Brooklyn, New York
3.Washington-2200宾夕法尼亚大道
4.爱莫利维尔码头
5.Woody Creek Garden, Pitkin County, Colorado
6.爱莫利维尔码头

Chair ○ 座椅

3

4

5

6

1-3.澳大利亚悉尼-Ballast Point 公园
4.澳大利亚悉尼-Bondi 至Bronte 滨海走廊
5.澳大利亚悉尼-Foley公园
6-8.澳大利亚悉尼-Victoria Park Public Domain

Chair ○ 座椅

Chair ○ 座椅

2.澳大利亚悉尼-悉尼5号湿地
澳大利亚悉尼-Pimelea Play Grounds Western
Sydney Parklands
5-6.芭堤雅希尔顿酒店

1-4.芭堤雅希尔顿酒店
5,6.白香果
7.百家湖

Chair ○ 座椅

Chair ○ 座椅

1-4.柏林-Südkreuz火车站
5.半岛湖边小屋

1，2.半岛湖边小屋
3.傍花村温泉度假村
4.鲍恩海滨
5.北京-燕西台
6.北京-波特兰
7.北京-翠湖——教授的乐园

Chair ○ 座椅

Chair ○ 座椅

1-3.北京-过山车
4,5.北京-花石间
6,7.北京-纳帕尔湾

Chair ○ 座椅

.北京－京都高尔夫中式别墅
.北京－龙湾
.北京市－竹溪园
.北京市－西山美墅馆
，6.北京－泰禾运河岸上的院子

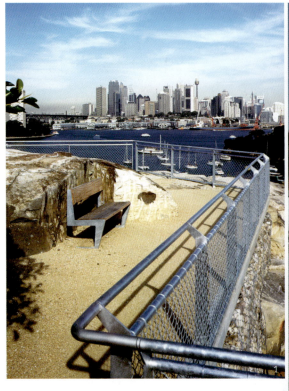

1.北悉尼地区-BP公司遗址公园
2.比利时Kapellen-Meeting square Kapellen
3.北京-天竺新新家园
4，5.北京-新新家园-庭院澜3

Chair ○ 座椅

1.北京-御墅临枫
2.财富公馆
3.北京圆明园-左右间咖啡的院
4,5.波士顿-帕梅特谷
6,7.碧湖别墅

Chair ○ 座椅

Chair ○ 座椅

1.波特兰花园60号
2.波特兰花园72号
3-5.布卡克海滩
6-8.成都美景金山

1.成都双流-和贵馨城
2-4.大湖山庄
5,6.达克兰的维多利亚港

Chair ○ 座椅

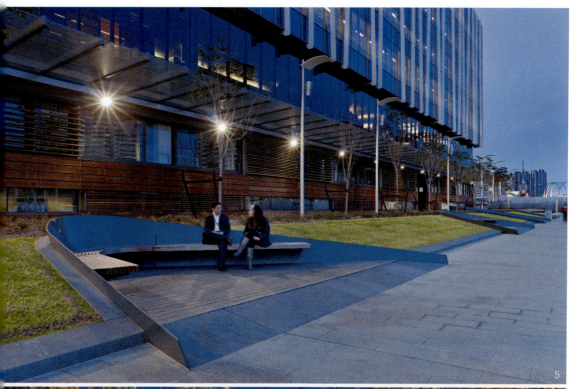

1.袋鼠湾
2.成都－玉都别墅庭园
3.丹麦，奥尔堡-DANISH CROWN-AREAS, NR. SUNDBY BRYGGE
4-7.丹麦-哥本哈根的西北公园

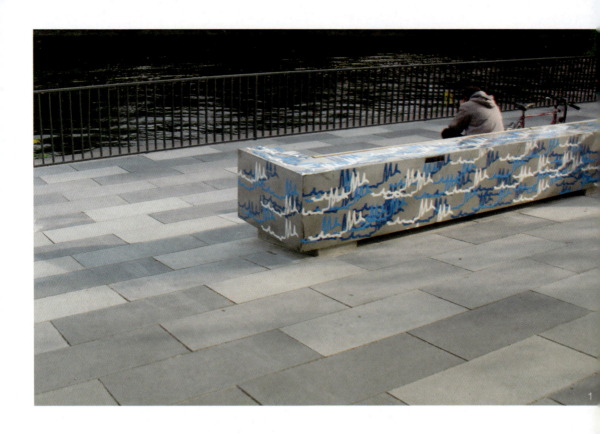

1，2.德国 柏林-南"制革厂"岛
3.德国巴伐利亚02007 Waldkirchen花园展
4-6.德国柏林-Imchen square Berlin-Kladow Redesign of square and waterfront promenade
7.德国柏林-国家展会（ULAP）广场
8.德国柏林-晋城市儿童公园

Chair ○ 座椅

Chair ○ 座椅

1-4.德国-什未林国家花园展
5,6.东方普罗旺斯（欧式）

155

Chair ○ 座椅

1,2.东方普罗旺斯（欧式）
3.东莞市-湖景壹号庄园私家庭院
4.俄勒冈州俄勒冈市-乔恩斯托姆公园
5.都柏林-倾斜空间-gal 4
6.都柏林-诺曼底-NORMANDIE. Design by Hugh Ryan. Photo by H Ryan

Chair ○ 座椅

1.翡翠岛
2-4.法国克洛尔-Renewal of the main street
5.观塘别墅庭院
6.佛山-天安鸿基花园

1.复地朗香80-1
2.广州-华南碧桂园
3.韩国汉城-ChonGae Canal Restoration Project
4.广东省广州市-力迅·上筑
5.韩国首尔-West Seoul Lake Park
6.河盛城邦
7.河源城区-河源东江·首府花园
8.荷兰Cuijk-Maaskade Cuijk
9.荷兰阿姆斯特丹-"Meer-park" Amsterdam

Chair ○ 座椅

161

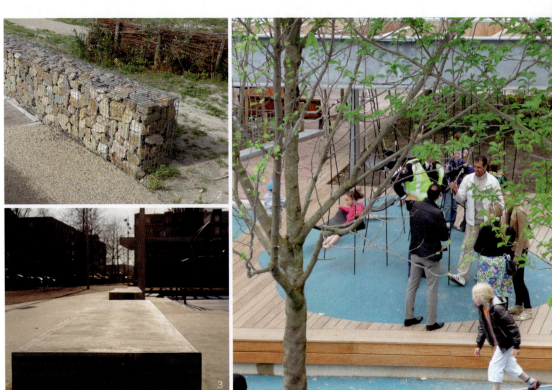

Chair ○ 座椅

1. 荷兰阿姆斯特丹－"Meer-park" Amsterdam
2. 荷兰阿姆斯特丹－Floriande居住区
3，4.荷兰阿姆斯特丹－van Beuningenplein
5. 荷兰海牙－playground Melis Stokepark
6. 荷兰海牙－square Beukplein
7，8.红木林别墅花园nEO

1，2.华盛顿-互惠中心屋顶花园
3.湖景壹号别墅庄
4.华南碧桂园燕园
5，6.华盛顿-水桌庭园
7，8.皇后区-MTA雨洪减滞装置

Chair ○ 座椅

1.加利福尼亚-特伦顿大道
2.加利福尼亚圣塔莫妮卡-Euclid Park
3-7.加拿大-东南福溪可持续社区

Chair ○ 座椅

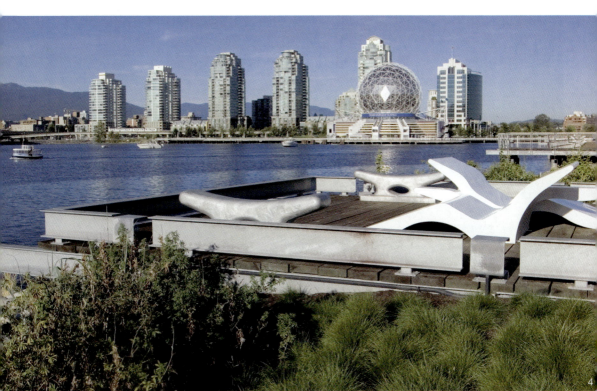

1,2.加拿大多伦多-Sugar Beach
3.加拿大蒙特利尔-Square Dorchester—Place du Canada
4-7.杰克·埃文斯船港—堤维德岬一期工程

Chair ○ 座椅

Chair ○ 座椅

1.兰桥圣菲91号
2，3.乐山市-恒邦·翡翠国际社区
4.里昂市-里昂河岸
5.罗德岛普罗维登斯 – Dunkin Donuts Plaza – Horizon Garden
6.洛杉矶 – 南加州大学医学中心

Chair ○ 座椅

1，2.洛杉矶－南加州大学医学中心RCH_LAC USC_Bonner_5489_014C
3.洛杉－加州威尼斯612_06
4.曼萨纳雷斯线性公园avp2©anamuller_300
5.美国波特兰－街区
6，7.美国俄勒冈－Jon Storm Park

Chair ○ 座椅

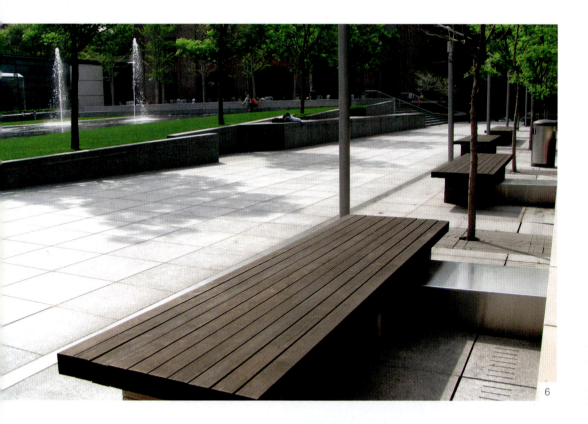

1.美国-哥伦比亚广场
2.美国加利福尼亚-特伦顿大道
3.美国马萨诸塞州波士顿-Levinson Plaza, Mission Park
4，5.美国纽约-Carnegie Hill House
6，7.美国纽约-亚瑟·罗斯平台
8.墨尔本-Elwood海滩景观设计

1，2.墨尔本-Elwood海滩景观设计
3.梦归地中海
4，5.明尼阿波利斯-史拜克曼住宅景观
6-8.墨西哥-城市与环境设计事务所

Chair ○ 座椅

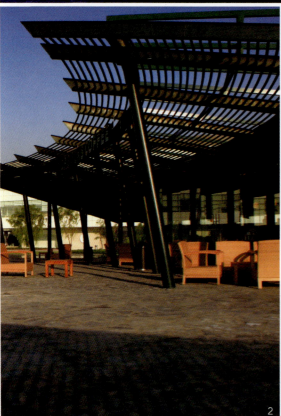

Chair ○ 座椅

1-3.墨西哥－高科技办公园区2.
4.墨西哥－马里纳尔可住宅136-04
5，6.慕尼黑－施瓦宾花园城市
7.纽约-龙门广场州立公园

Chair ○ 座椅

1-3.挪威奥斯陆-Rolfsbukta住宅区
4-6.挪威奥斯陆-Tjuvholmen
7.挪威-南森公园
8.盘龙城

1，2.泉州－奥林匹克花园别墅区67高挡土墙与休闲平台的自然处理
3.日本茨城县-福冈堰樱花公园
4-6.日式枯山水庭院
7，8.瑞士日内瓦-Renewal of Place des Nations
9.瑞士-Enea景观设计公司总部

Chair ○ 座椅

183

Chair ○ 座椅

1,2.瑞士-Enea景观设计公司总部
3.萨缪尔·德·尚普兰滨水长廊
4.上海-别墅花园
5-7.上海-兰乔圣菲
8,9.上海-丽茵别墅

1.上海-玫瑰园
2，3.上海-仁恒屋顶花园
4，5.上海-圣安德鲁斯庄园
6.上海市-比华利别墅
7.上海市-比利华2
8.上海市-春天花园

Chair ○ 座椅

1，2.上海市-观庭
3，4.上海市-豪嘉府邸别墅花园
5.上海市 - 圣塔路斯别墅花园
6.上海市-太阳湖大花园设计造园
7，8.上海市-花语墅

Chair ○ 座椅

1-3.上海市-汤臣高尔夫别墅
4,5.上海-太阳湖大花园设计造园
6,7.上海-太原路小别墅
8,9.上海-田园别墅花园

Chair ○ 座椅

4

5

6

7

8

9

1.上海－万科燕南园
2，3.上海－西郊大公馆
4-6.上海－夏州花园
7.上海－新律花园
8.上海－小别墅

Chair ○ 座椅

4

5

6

7

8

1.Anchorage, Alaska, USA-Anchorage Museum Expansion
2.佘山-3号园
3.佘山-东紫园
4.佘山的梦想
5,6.佘山-东紫园

1-6.佘山－高尔夫别墅花园

Chair ○ 座椅

1，2.佘山－高尔夫别墅花园
3.深圳-SHENZHEN BAY COASTLINE PARK
4.深圳－万科金域华府
5.雪梨澳乡别墅庭院
6.新西兰-杰利科北部码头漫步长廊，杰利科大道和筒仓公园
7.亚德韦谢姆大屠杀博物馆大楼
8.亚利桑那州－理工学院景观

Chair ○ 座椅

Chair ○ 座椅

1.深圳－卧式摩天楼IMG_1713-med
2，3.圣安东尼奥德克萨斯-Main Plaza Shade Structures
4.圣达菲铁路站场公园
5，6.圣地亚哥-拉卡斯塔格伦社区

1,2.台湾-基隆海洋广场
3,4.台湾-南投农舍别墅-backyard

Chair ○ 座椅

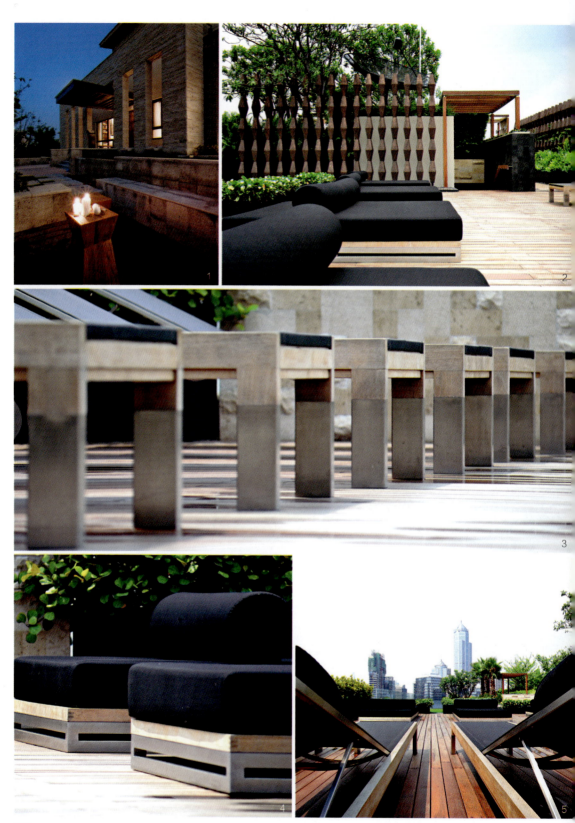

Chair ○ 座椅

1.太湖高尔夫
2-5.泰国曼谷-Prive by Sansiri
6-8.唐山－唐人起居

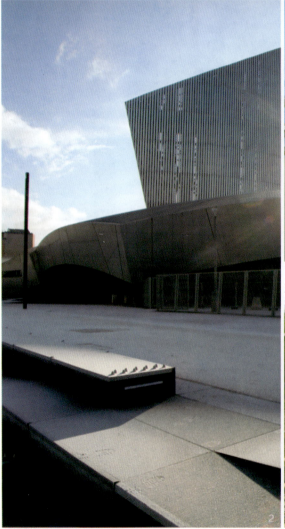

Chair ○ 座椅

1. 唐山－唐人起居
2，3. 特拉福德滨河长廊
4. 提香草堂4
5-7. 同润加州

1,2.天鹅湖
3.万成华府
4,5.万科－金域华府
6,7.万科－兰乔圣菲

Chair ○ 座椅

Chair ○ 座椅

1，2.万科－兰乔圣菲
3.万科朗润园
4.挪威奥斯陆－Tjuvholmen
5，6.威特市斯博拉茨
7，8.天津－桥园公园

1，2.威特市斯博拉茨
3，4.文华别墅
5.五栋大楼屋顶花园
6，7.五溪御龙湾
8，9.西班牙伊伦市-Alai Txoko公园

Chair ○ 座椅

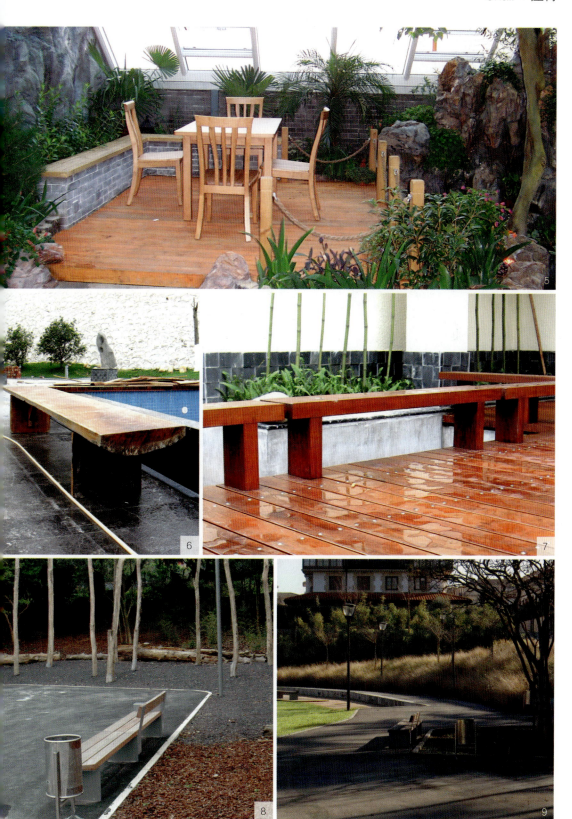

Chair ○ 座椅

1.2.悉尼-Pirrama公园
3-5.悉尼-奥林匹克公园Jacaranda广场
6.悉尼－海洋生物站公园
7.新南威尔士帕丁顿-PADDINGTON RESERVOIR GARDENS

215

1. 亚利桑那州-理工学院景观_ASU Poly 07
2. 亚利桑那州-梅萨艺术中心207-14
3. 亚利桑那州-石英山居所,534-02
4. 加拿大-东南福溪可持续社区
5. 亚洲大酒店
6. 阳光九九造园
7,8. 叶片小斋leaf house

Chair ○ 座椅

Chair ○ 座椅

1，2.以色列贝尔谢巴-BGU University Entrance Square && Art Gallery
3.以色列哈尔阿达尔-Har Adar #1 住宅
4.以色列-海尔兹利亚公园
5，6.以色列-特拉维夫港口
7.易郡20090708020

Chair ○ 座椅

1.意大利卢卡-圣·安娜区公园
2.意大利洛迪-漫步花园
3.银都名墅
4.云栖蝶谷
5，6.张家港 怡佳苑

1. 重庆－梦中天地
2. 上海市－汤臣高尔夫别墅
3. 钻石提格公园

ARDEN LANDSCAPE DETAIL DESIGN · SIGN BOARD

标志牌

　　庭院的意义在于居住者对它所提供的生活方式的认同，在于它所带来的院落"能指"（居住环境）和"所指"（居住者的生活方式、社会交往方式）的和谐关系，而此代码在现代集合住宅中的被扭曲和遗忘引发了居民对院落中亲密的邻里关系的怀念。

　　庭院作为人为的自然空间，它将自然和人工相结合起来，让人产生一种回归自然的向往。庭院中往往常采用小巧、精致的手法，设置植物、院路、亭廊、假山、雕塑、水池等，既表现出个性化的特点，同时给人们带来自然美、人工美的艺术享受。

gnatus eiumendae voluptamusa custet et demporeribus minverchil mi, unte magnit eum et que doloreh enihili uaspe vendi ium nume sum vita doloria nosam as utendi te solori beribus pa consequi idias maiorem porestiis xerendias as es ut volupit, el earum labo. Itatemp orerenes ipsunt, simusap idustenet explacea et.
on rat doluptatus arum et que plit occum et iunt quam, voluptia parcius dollect aturehentum abo. Neque voluptas eque sam, qui opta volorib ernate lat latem eumet la quis qui ut arit, am reperates dolore.

Sign Board ○ 标志牌

1，2.Adelaide Zoo Giant Panda Forest
3.Anchorage, Alaska, USA–Anchorage Museum Expansion
4.Cairnlea, Victoria–Cairnlea
5.Fredericia‐Temporary Park
6.Genk Belgium–m!ne
7.Santa Monica–欧几里德公园–10 RCH_Euclid Park_Bonner_5481_008C
8，9.Eugene, Oregon, USA–John E. Jaqua Academic Center for Students Athletes

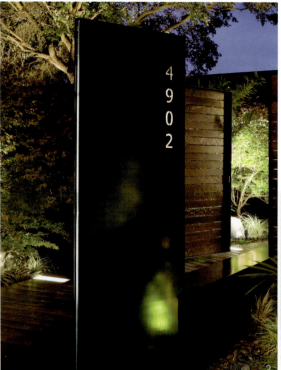

1.Toronto, Ontario, Canada-HtO
2.半岛湖边小屋
3.安徽省-合肥政务文化主题公园
4.Unfolding Terrace, Dumbo, Brooklyn, New York
5，6.荷兰阿姆斯特丹-"Meer-park" Amsterdam

Sign Board 标志牌

1.德国柏林-莫阿比特监狱历史公园
2.广东东莞-东莞长城世家
3.广州-保利国际广场
4.瑞士日内瓦-Renewal of Place des Nations
5.汇景新城

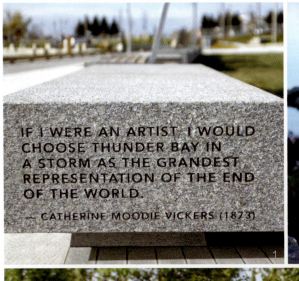

1,2.桑德贝滨水景观,亚瑟王子码头
3.上海－万科深蓝
4.上海市－东町庭院
5.佘山－东紫园
6.上海市－汤臣高尔夫别墅
7.加拿大,私人公寓,兰乔圣菲
8.上海市－御翠园

Sign Board ○ 标志牌

Sign Board ○ 标志牌

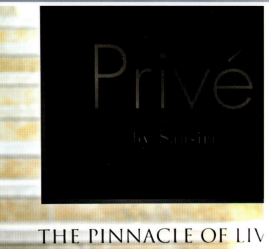

1. 威特市斯博拉茨
2. 深圳－万科金域华府－现代艺术意味的主入口
3，4. 台湾－基隆海洋广场
5，6. 泰国曼谷－Prive by Sansiri

1-4.新南威尔士帕丁顿-PADDINGTON RESERVOIR GARDENS

GARDEN LANDSCAPE DETAIL DESIGN · SCULPTURE SKETCH

雕塑小品

　　庭院的意义在于居住者对它所提供的生活方式的认同，在于它所带来的院落"能指"（居住环境）和"所指"（居住者的生活方式、社会交往方式）的和谐关系，而此代码在现代集合住宅中的被扭曲和遗忘引发了居民对院落中亲密的邻里关系的怀念。

　　庭院作为人为的自然空间，它将自然和人工相结合起来，让人产生一种回归自然的向往。庭院中往往常采用小巧、精致的手法，设置植物、院路、亭廊、假山、雕塑、水池等，既表现出个性化的特点，同时给人们带来自然美、人工美的艺术享受。

gnatus eiumendae voluptamusa custet et demporeribus minverchil mi, unte magnit eum et que doloreh enihili uaspe vendi ium nume sum vita doloria nosam as utendi te solori beribus pa consequi idias maiorem porestiis xerendias as es ut volupit, el earum labo. Itatemp orerenes ipsunt, simusap idustenet explacea et. on rat doluptatus arum et que plit occum et iunt quam, voluptia parcius dollect aturehentum abo. Neque voluptas eque sam, qui opta volorib ernate lat latem eumet la quis qui ut arit, am reperates dolore.

Sculpture sketch ○ 雕塑小品

1-3.Aalborg, Denmark-奥尔堡滨水 - 港口与城市连接
4，5.Adelaide Zoo Giant Panda Forest
6.Alameda, USA Date of Completion-海湾船艇公司

Sculpture sketch ○ 雕塑小品

1，2.Alameda, USA Date of Completion－海湾船艇公司
3.Eugene, Oregon, USA－John E. Jaqua Academic Center for Students Athletes
4.Erman Residence, San Francisco, California
5.Houston, TX－ConocoPhillips World Headquarters
6，7.Houston, TX－The Brochstein Pavilion
8.Alameda, USA Date of Completion－海湾船艇公司
9.New York－南部河畔公园
10.Il nuovo Parco di Jean Nouvel a Barcellona

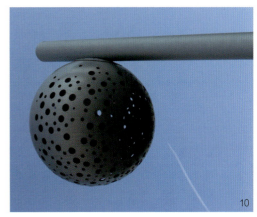

Sculpture sketch ○ 雕塑小品

1-4.London－水域郊野公园游乐区
5.Brumadinho, Brazil Date of Completion－Burle Marx教育中心
6,7.Jeonnam－小岛上的动物园－JDS_DOCHODO ISLAND ZOO－AVIARY CLOSE UP
8.Lakeway Drive － In Redevlopment

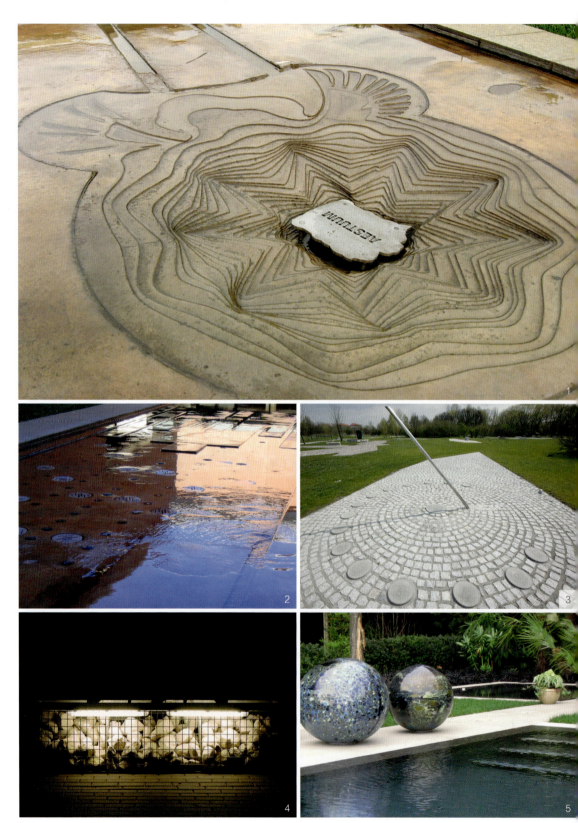

Sculpture sketch ○ 雕塑小品

1，2.South Africa-University of Johannesburg Arts Centre
3，4.Stanislaw Lem memorial - Garden of Experience - Educational Park in Krakow
5-7.Tables of Water, Lake Washington, Washington
8.The Hague, The Netherlands-park Vlaskamp

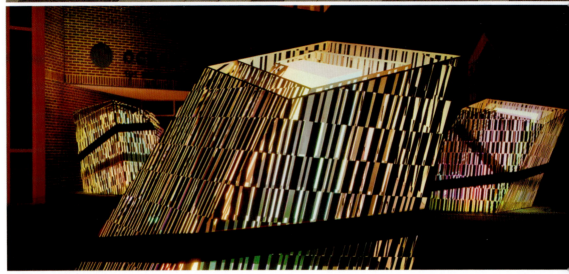

Sculpture sketch ○ 雕塑小品

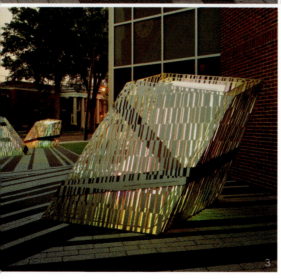

1-3.Toms River, New Jersey-Ocean County Public Library – BarCODE Luminescence
4.Santa Monica-欧几里德公园-06 RCH_Euclid Park_Bonner_5481_006C
5.San Luis Potosí, Mexico Date of Completion-圣路易总体规划

1,2.爱尔兰-"自我"圣坛
3,4.爱尔兰都柏林-红杉（展示花园）
5,6.阿姆斯特丹-辛克尔小岛
7-10.Tolosa (Gipúzcoa)-Jolastoki (tolosa)

Sculpture sketch ○ 雕塑小品

1.澳大利亚悉尼-Ballast Point 公园
2.白香果
3.澳大利亚悉尼-Pimelea Play Grounds Western Sydney Parklands
4.澳大利亚悉尼-Bondi 至 Bronte 滨海走廊
5.澳大利亚-东联高速公路
6.澳大利亚-库吉港
7-9.芭堤雅希尔顿酒店

Sculpture sketch ○ 雕塑小品

Sculpture sketch ○ 雕塑小品

傍花村温泉度假村
澳大利亚-库吉港
，4.北京-燕西台
北京-过山车
爱尔兰都柏林-观海

1.北京市-润泽庄园2
2.北京市-竹溪园
3.北京-泰禾运河岸上的院子
4，5.北京-天竺新新家园
6.北京-新新家园
7，8.北京-易郡别墅

Sculpture sketch ○ 雕塑小品

1，4.布卡克海滩
2，3，5.财富公馆
6.成都双流-和贵馨城

Sculpture sketch ○ 雕塑小品

255

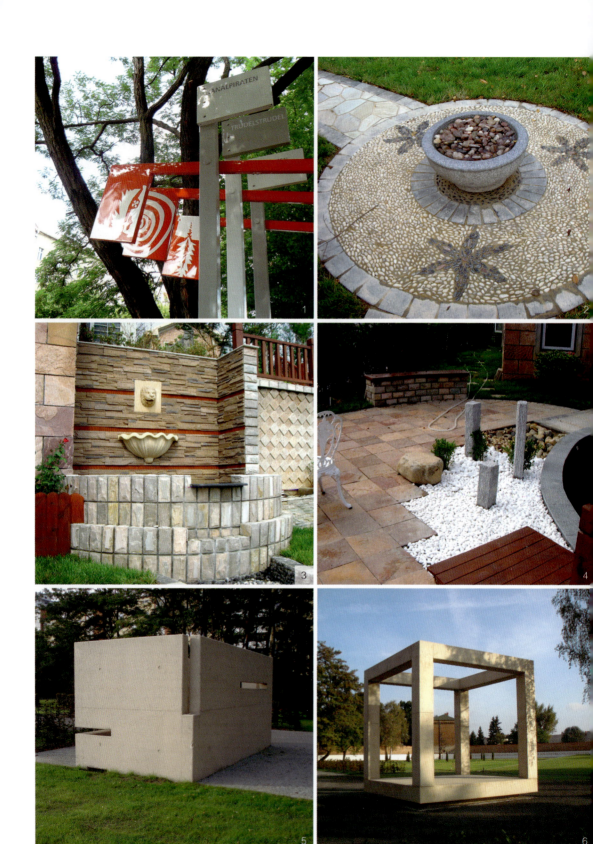

Sculpture sketch ○ 雕塑小品

1. 德国 柏林-南"制革厂"岛
2-4. 帝景天成坡地别墅
5，6. 德国柏林-莫阿比特监狱历史公园
7. 德国巴伐利亚02007 Waldkirchen花园展
8-10. 德国巴伐利亚州-德累斯顿动物园-长颈鹿园
11. 德国柏林-莫阿比特监狱历史公园

1.东莞市-湖景壹号庄园私家庭
2，3.东山墅
4.俄勒冈州俄勒冈市-乔恩斯托姆公园
5.法国波尔多-Le miroir d'eau
6，7.法国克洛尔-Renewal of the main street

Sculpture sketch ○ 雕塑小品

Sculpture sketch ○ 雕塑小品

.北京-过山车
, 3.北京-花石间
.北京 - 龙湾2
.北京-龙湾别墅和院
, 7.北京 - 润泽庄园

1. 芬洛市-马斯河大街—芬洛马斯河岸新城区
2-4. 佛山－天安鸿基花园
5-7. 复地朗香75-4

Sculpture sketch ○ 雕塑小品

Sculpture sketch ○ 雕塑小品

1-4.广东东莞-东莞长城世家
5.广州-凤凰城8号
6.广州-汇景新城别墅
7.广州-金碧花园
8.桂林南溪宾馆
9.河盛城邦

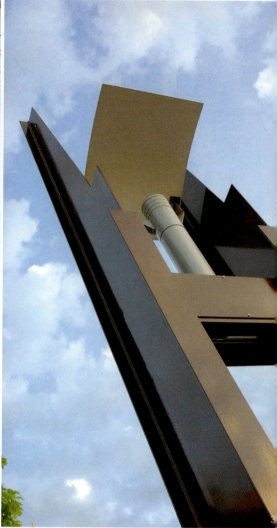

1,2.荷兰Getsewoud Zuid-schoolyard and playground Getsewoud Zuid
3.加拿大-东南福溪可持续社区
4.荷兰Heerhugowaard-South-Park of Luna
5,6.荷兰Velsen-public park Duinpark

Sculpture sketch ○ 雕塑小品

1.成都－玉都别墅庭园
2.达尔文海滨公共区域
3，4.丹麦-哥本哈根的西北公园
5.荷兰阿姆斯特丹－"Meer-park" Amsterdam
6.荷兰阿姆斯特丹-van Beuningenplein

Sculpture sketch ○ 雕塑小品

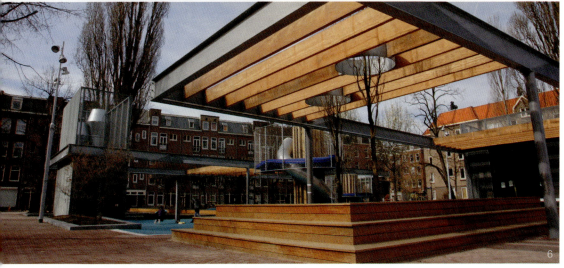

Sculpture sketch ○ 雕塑小品

1,2.荷兰海牙-Palace Gardens
3,4.荷兰海牙-playground Melis Stokepark
5.荷兰海牙-square Beukplein

Sculpture sketch ○ 雕塑小品

1，2.湖景壹号别墅庄
3.红木林别墅花园nEO
4，5.华南碧桂园燕园
6.华盛顿-互惠中心屋顶花园

1-3.加拿大多伦多-Sugar Beach
4-6.加拿大魁北克-Core Sample
7.加拿大蒙特利尔-Square Dorchester—Place du Canada

Sculpture sketch ○ 雕塑小品

1.昆明－中美二战友谊公园07_011室外展场效果图
2.兰阿姆斯特丹-public square Columbusplein
3.兰桥圣菲91号
4.美国波特兰-街区
5.乐山市-恒邦·翡翠国际社区
6-8.伦敦－现代茶园洋房－Knightsbridge
9.美国-哥伦比亚广场

Sculpture sketch ○ 雕塑小品

1,2.绿洲千岛花园
3.美国加利福尼亚-特伦顿大道
4-6.美国旧金山-Cow Hollow学校操场
7.美国纽约-亚瑟·罗斯平台
8.美国普罗维登斯-The Steel Yard

Sculpture sketch ○ 雕塑小品

1，2.梦归地中海
3.荷兰阿姆斯特丹-"Meer-park" Amsterdam
4，5.墨尔本-儿童艺术游乐园
6.墨尔本-Elwood海滩景观设计
7.荷兰Velsen-public park Duinpark

Sculpture sketch ○ 雕塑小品

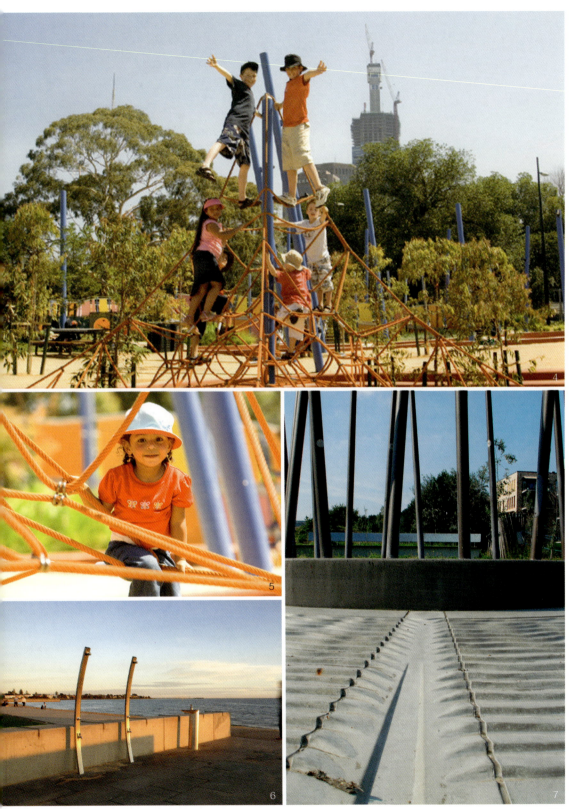

1，2.墨尔本-儿童艺术游乐园
3，4.墨西哥墨西哥城-Radial Garden Mexico Historical Center
5.西班牙伊伦市-Alai Txoko公园
6.慕尼黑-施瓦宾花园城市

Sculpture sketch ○ 雕塑小品

1,2.挪威奥斯陆-Pilestredet公园
3,4.挪威奥斯陆-Tjuvholmen
5,6.挪威-南森公园
7.盘龙城

Sculpture sketch ○ 雕塑小品

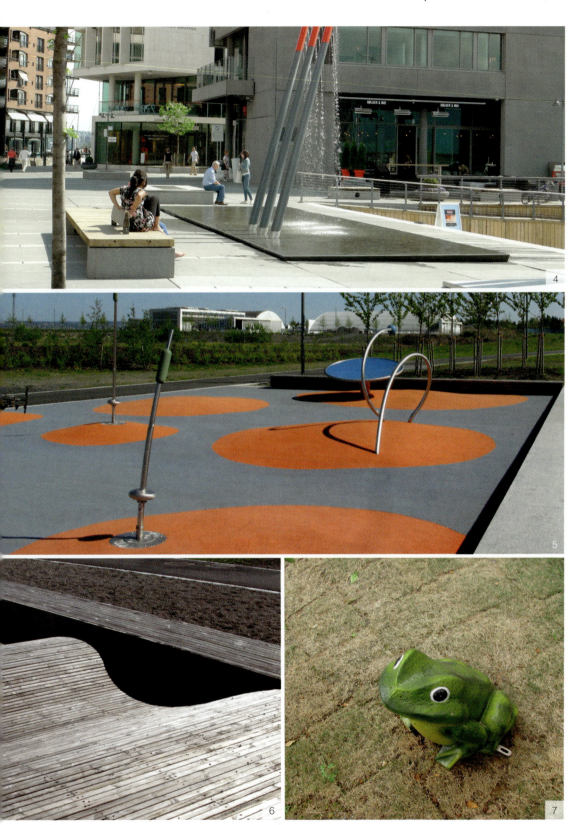

1，2.盘龙城
3.秦皇岛－海滩修复工程
4.墨尔本－Frankston
5，6.泉州－奥林匹克花园别墅区

Sculpture sketch ○ 雕塑小品

4

5

6

1-3.日式枯山水庭院
4.瑞士日内瓦-Renewal of Place des Nations
5-7.桑德贝滨水景观,亚瑟王子码头

Sculpture sketch ○ 雕塑小品

Sculpture sketch ○ 雕塑小品

1，2.桑德贝滨水景观，亚瑟王子码头
3，4.上海－－辰山植物园
5.上海－底楼花园

Sculpture sketch ○ 雕塑小品

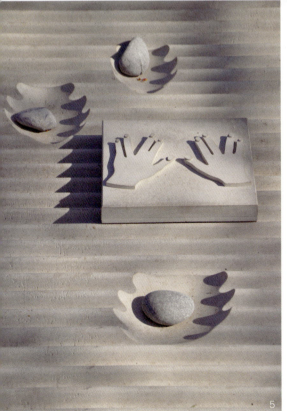

1-3.上海－绿洲江南园
4，5.宗教墓葬文化展示园FG
6..上海－玫瑰园

Sculpture sketch ○ 雕塑小品

1-3.上海－圣安德鲁斯庄园
4.上海市-Sarah的花园
5.上海市-比华利别墅
6-8.上海市-比华利1

1，2.上海市-比利华2
3.上海-万科深蓝
4-6.上海市-大华锦绣

Sculpture sketch ○ 雕塑小品

1.上海市-观庭
2.上海市-豪嘉府邸别墅花园
3-6.上海市-华庭雅居别墅花园
7.上海市-环球翡翠湾联排别墅

Sculpture sketch ○ 雕塑小品

Sculpture sketch ○ 雕塑小品

1.上海市-金地格林
2-4.上海市-佘山高尔夫夜景
5,6.上海市-圣塔路斯别墅花园
7.上海市-太阳湖大花园设计造园

1-3.上海市-太阳湖大花园设计造园
4.上海市－汤臣高尔夫别墅
5.上海市-御翠园
6.上海市-万科燕南园
7.上海市－汤臣高尔夫别墅
8 上海－太原路小别墅

Sculpture sketch ○ 雕塑小品

1.上海－太原路小别墅
2，3.上海－万科朗润园
4.上海－西郊大公馆
5-7.上海－万科深蓝

Sculpture sketch ○ 雕塑小品

1，2.深业.紫麟山
3.荷兰Velsen-public park Duinpark
4.深圳－东部华侨城湿地花园
5.深圳－万科金域华府
6.深圳－卧式摩天楼Entrance_DragonHead
7.深圳－中海大山地

Sculpture sketch ○ 雕塑小品

1-3.圣安东尼奥德克萨斯-Main Plaza Shade Structures
4.沈阳-万科金域蓝湾23
5，6.圣地亚哥-拉卡斯塔格伦社区

309

Sculpture sketch ○ 雕塑小品

1，2.台湾-基隆海洋广场
3，4.台湾-南投农舍别墅 - backyard
5.台湾商会
6.北京-过山车
7.泰国曼谷-Prive by Sansiri

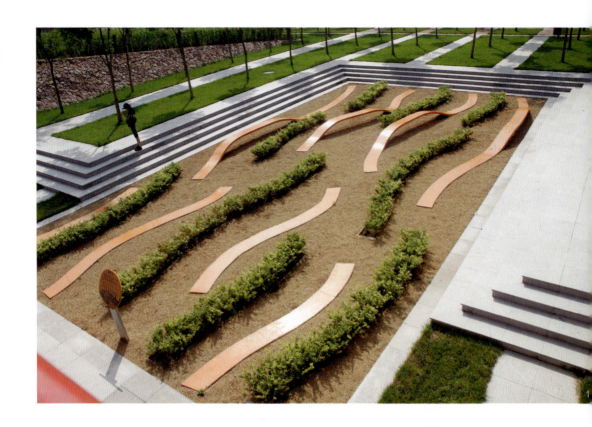

Sculpture sketch ○ 雕塑小品

1,2.天津－桥园公园
3.同润加州C
4.同润加州
5.万科朗润园
6.万科棠樾

1，2.西安-世园会之荷兰生态园-nEO
3.墨西哥-马里纳尔可住宅136-14
4-6.五溪御龙湾
7.武汉-天下别墅
8.西安 -世界园艺博览会-+scape!-flowing gardens-08

Sculpture sketch ○ 雕塑小品

1.亚利桑那州-理工学院景观
2.西郊一品
3.西雅图-山顶住宅
4-6.悉尼-Pirrama公园
7.新港码头
8.新加坡-City Square Urban Park

Sculpture sketch ○ 雕塑小品

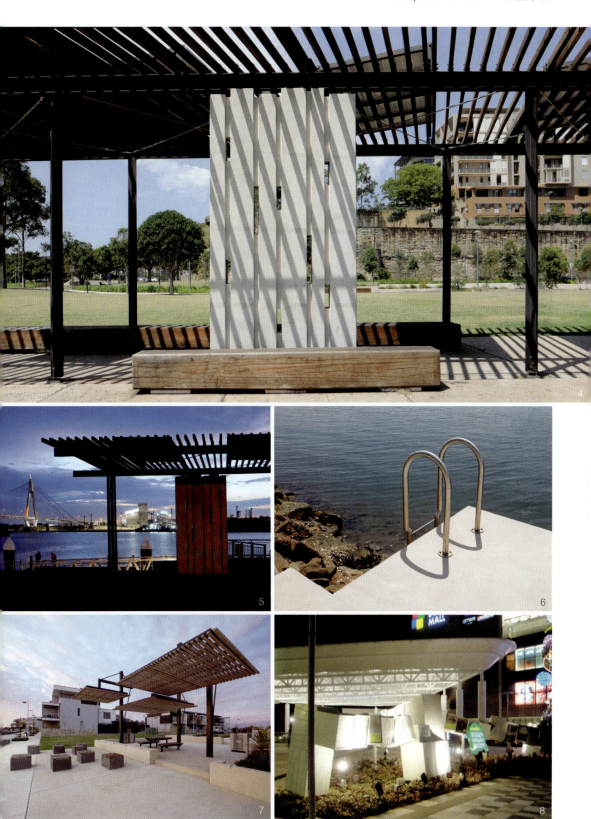

Sculpture sketch ○ 雕塑小品

1.依云听香阁后花园
2，3.以色列-海尔兹利亚公园
4.以色列贝尔谢巴-BGU University Entrance Square && Art Gallery
5，6.以色列-特拉维夫港口
7.易郡20090708009

1.钻石提格公园
2-4.宗教墓葬文化展示园FG